# PRE-K COUNT & COLOR

I0617238

1 2 3
4 5 6
7 8 9

Name: _____  Date: _____

# Color & Trace

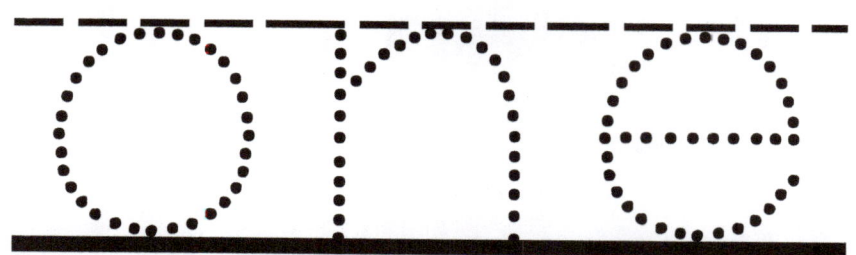

Nombre:_____ Fecha:_____

# Colorear y Trazar

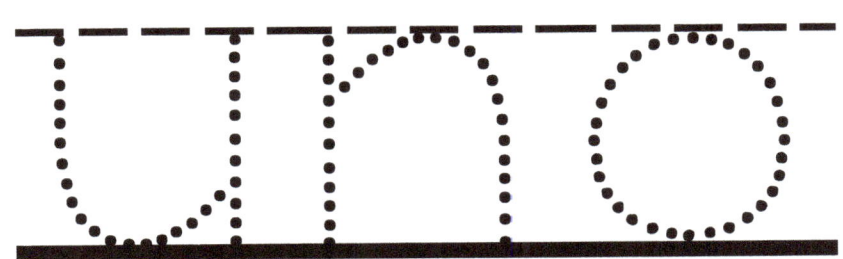

# Color & Trace

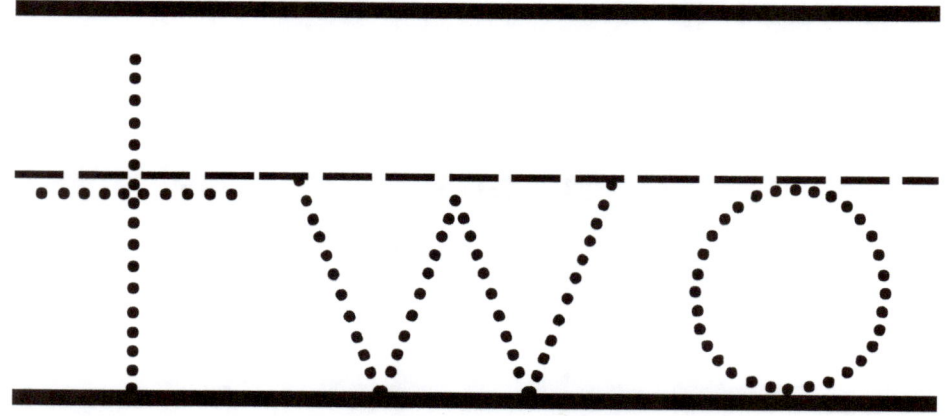

# Colorear y Trazar

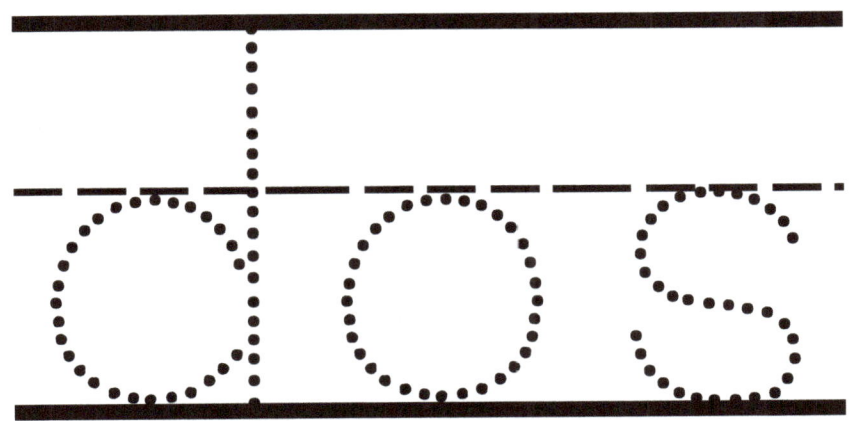

Name: _____ Date: _____

# Color & Trace

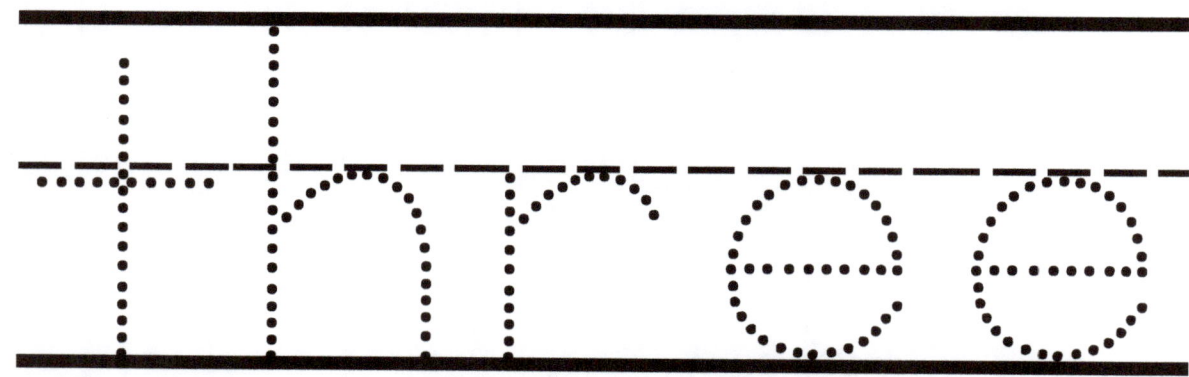

Nombre:_____ Fecha: _____

# Colorear y Trazar

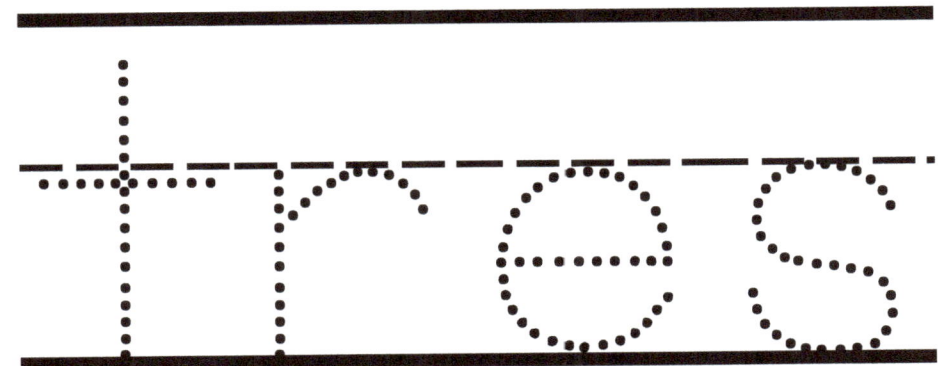

# Color & Trace

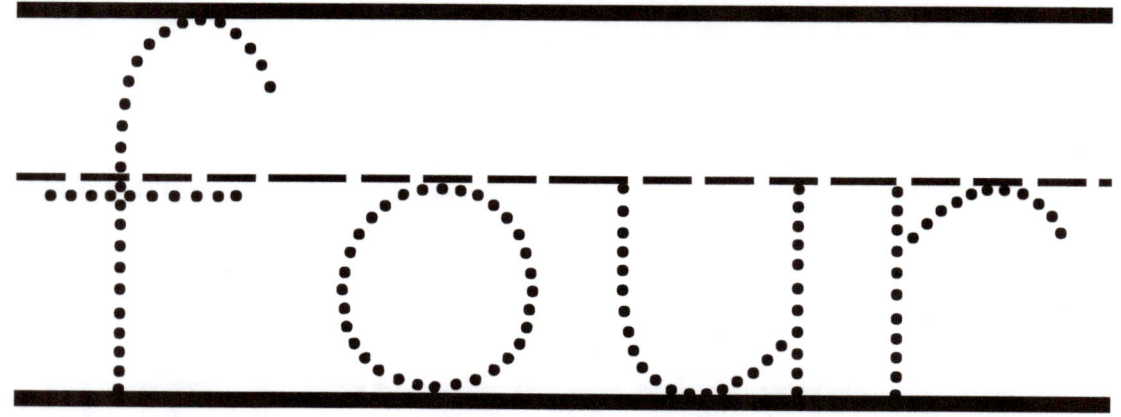

Nombre:_____ Fecha:_____

# Colorear y Trazar

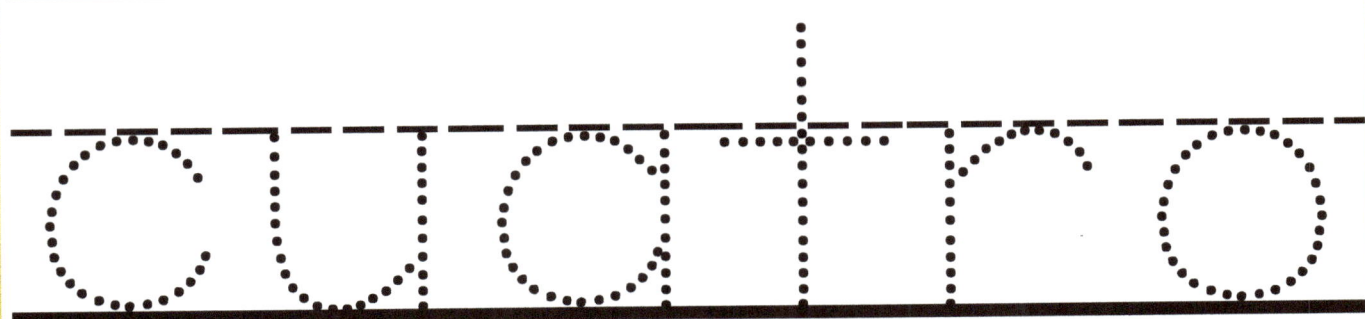

# Color & Trace

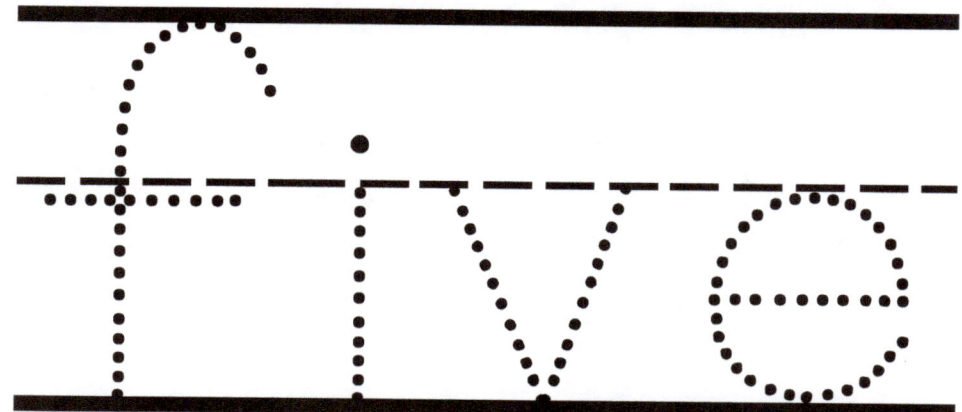

Nombre:_____ Fecha:_____

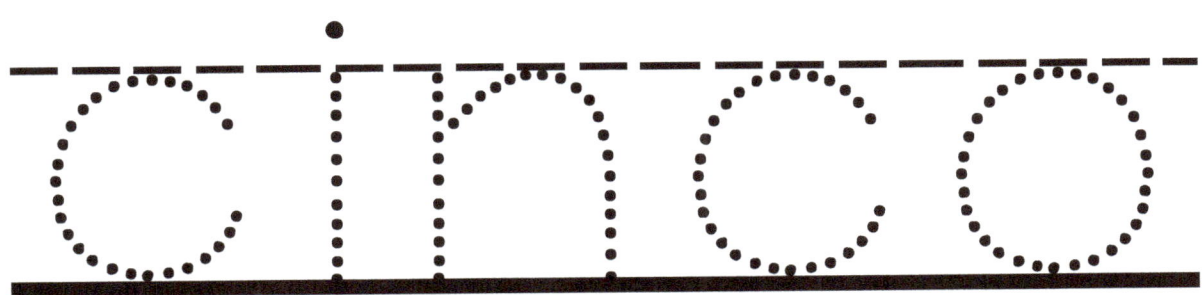

Name: _____ Date: _____

# Color & Trace

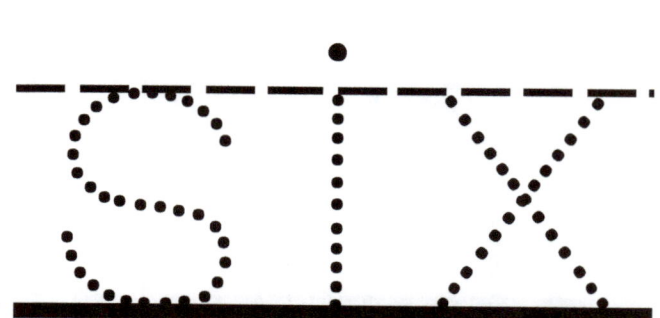

# Colorear y Trazar

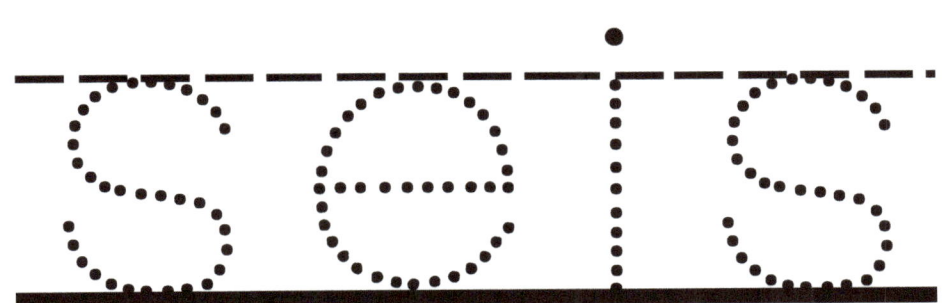

Name: _____ Date: _____

# Color & Trace

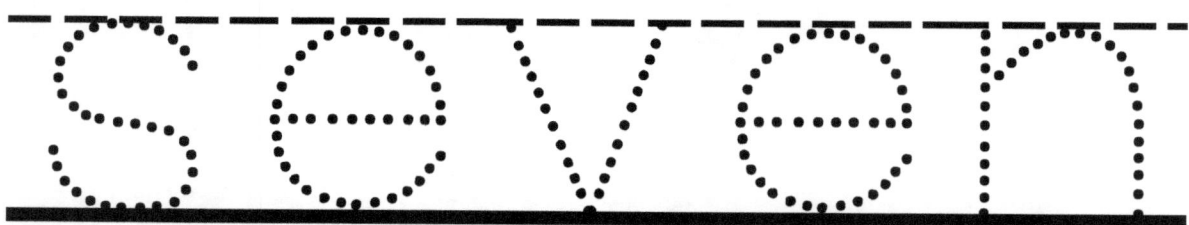

# Colorear y Trazar

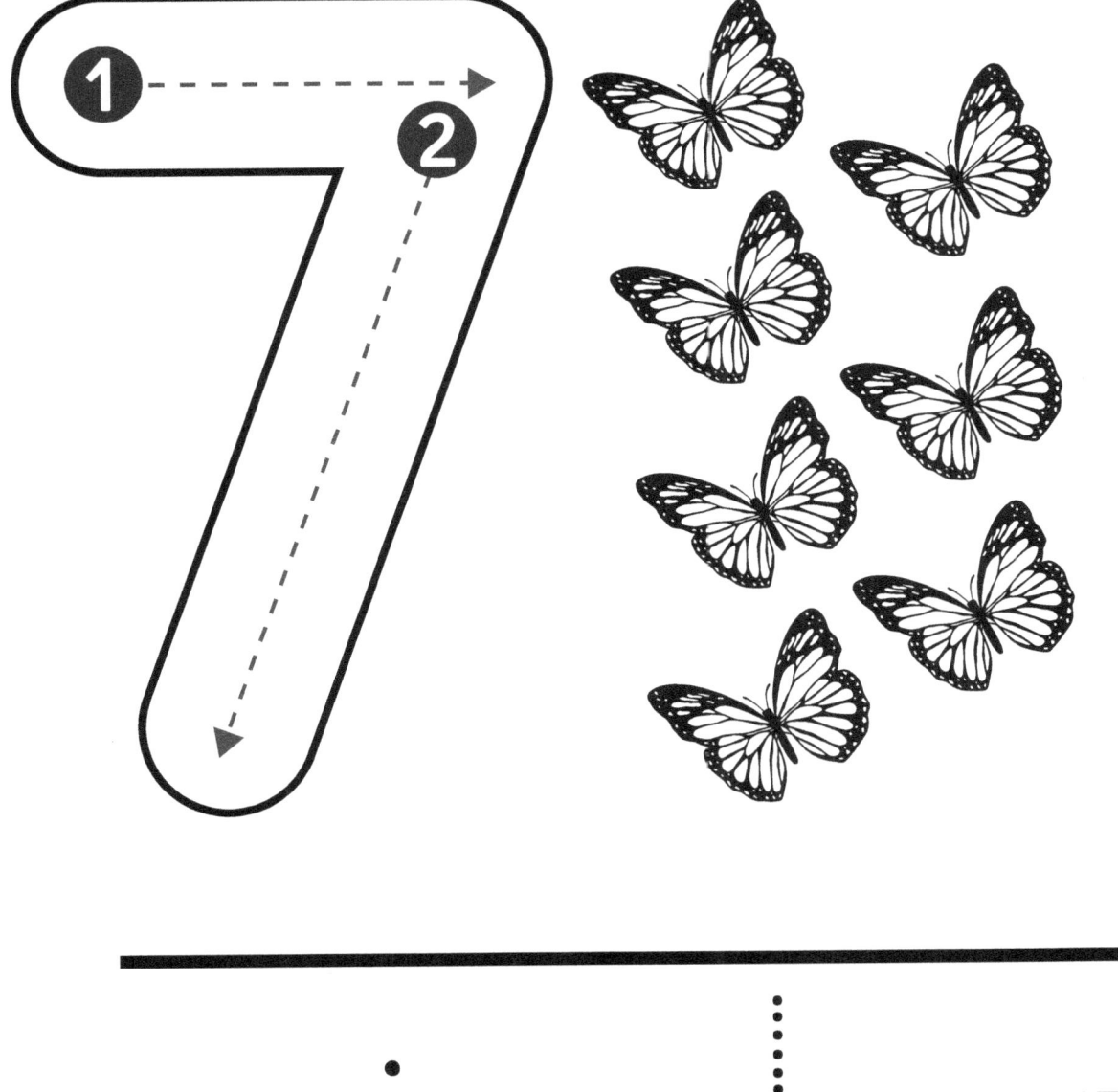

siete

**Name:** _____  **Date:** _____

# Color & Trace

8

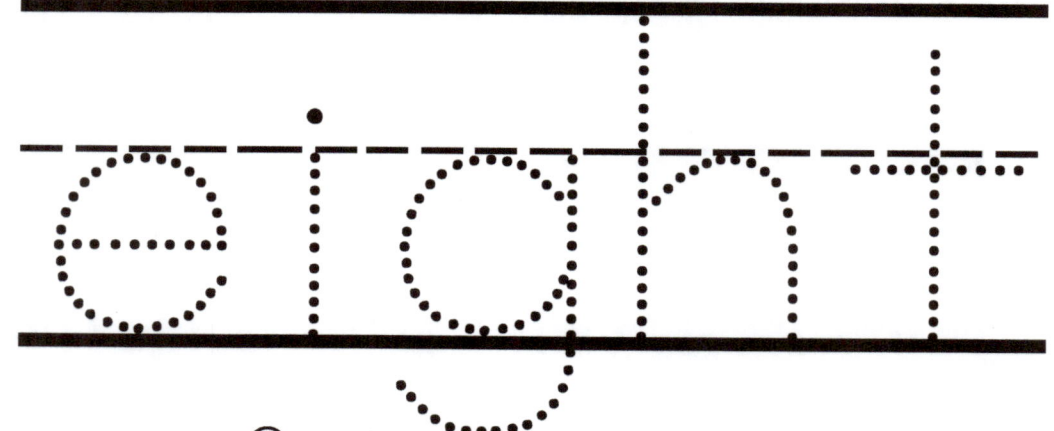

eight

Nombre:_____ Fecha:_____

# Colorear y Trazar

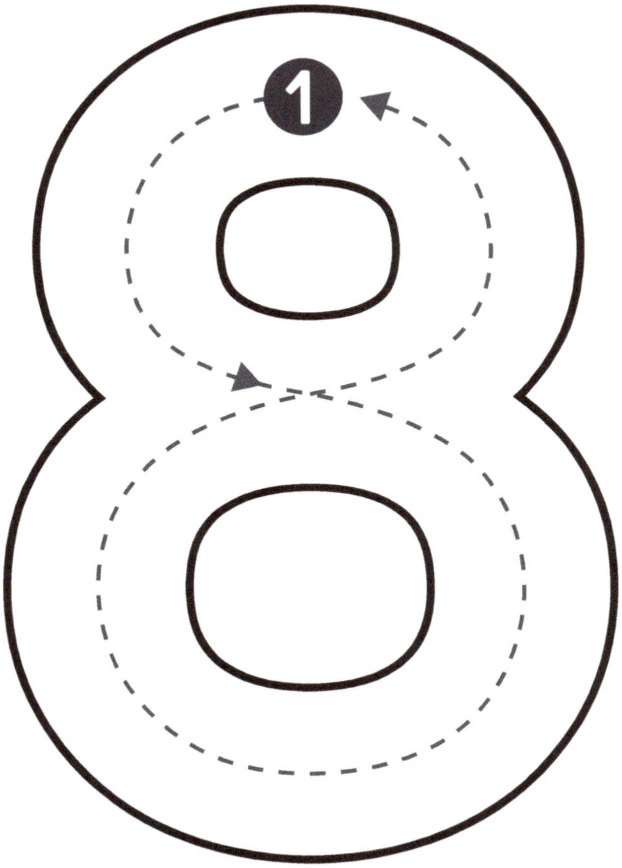

8

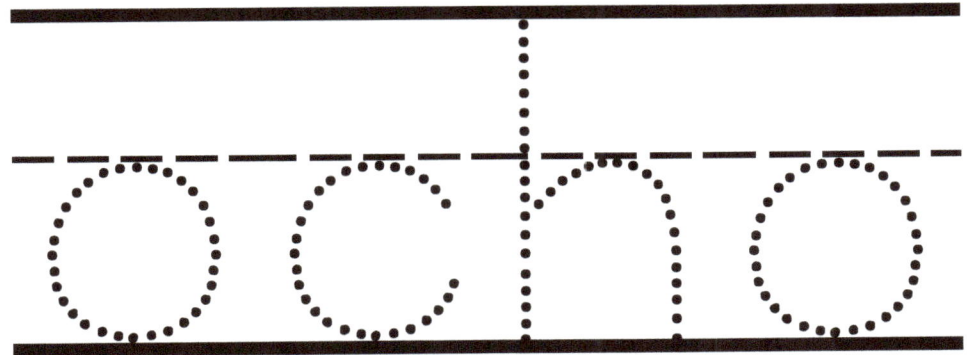

# Color & Trace

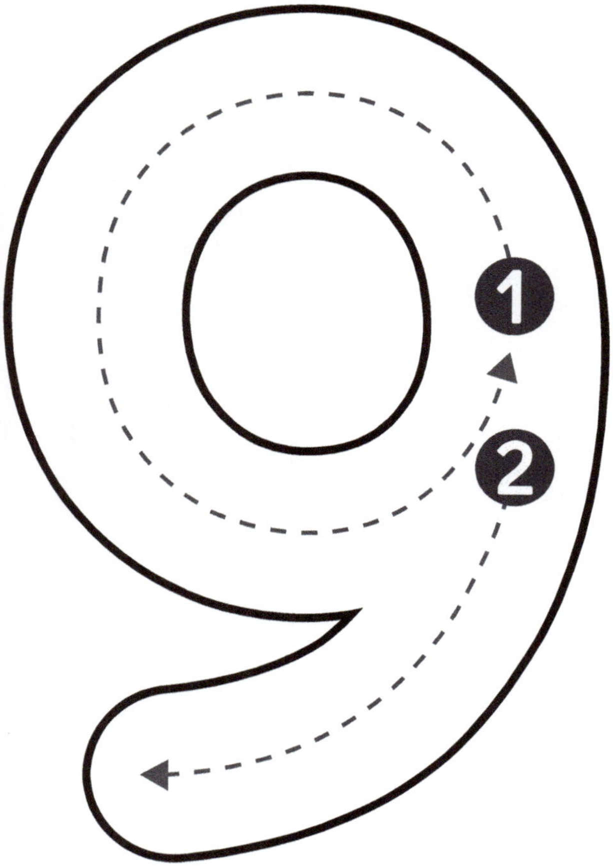

9
①
②

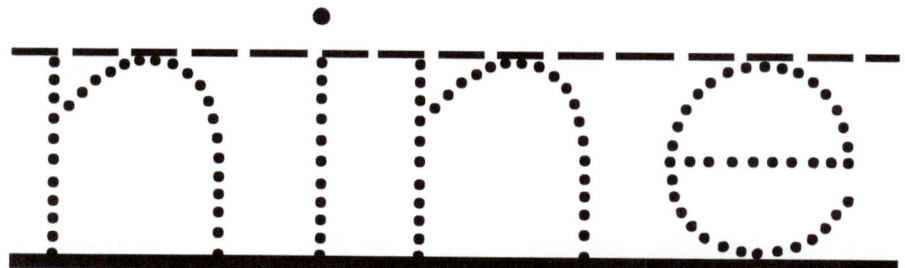

Nombre:_____ Fecha:_____

# Colorear y Trazar

nueve

**Name:** _____  **Date:** _____

# Color & Trace

Nombre:_____ Fecha: _____

# Colorear y Trazar

Name:_____ Date:_____

# TRACING NUMBERS 1-5

## Trace the numbers.

| 1 | 1 | 1 | 1 |
| 2 | 2 | 2 | 2 |
| 3 | 3 | 3 | 3 |
| 4 | 4 | 4 | 4 |
| 5 | 5 | 5 | 5 |

Nombre:_____ Fecha:_____

# TRAZAR LOS NUMEROS 1-5

## Trazar los números.

| | | | |
|---|---|---|---|
| 1 | 1 | 1 | 1 |
| 2 | 2 | 2 | 2 |
| 3 | 3 | 3 | 3 |
| 4 | 4 | 4 | 4 |
| 5 | 5 | 5 | 5 |

# TRACING NUMBERS 6-10

## Trace the numbers.

| | | | |
|---|---|---|---|
| 6 | 6 | 6 | 6 |
| 7 | 7 | 7 | 7 |
| 8 | 8 | 8 | 8 |
| 9 | 9 | 9 | 9 |
| 10 | 10 | 10 | 10 |

Nombre:_____ Fecha: _____

# TRAZAR LOS NUMEROS 6-10

## Trazar los números.

| 6 | 6 | 6 | 6 |
| 7 | 7 | 7 | 7 |
| 8 | 8 | 8 | 8 |
| 9 | 9 | 9 | 9 |
| 10 | 10 | 10 | 10 |

**Name:**_____ **Date:** _____

# MISSING NUMBERS 1-5

### Write the missing numbers.

| 1 | 2 | | 4 | 5 |
|---|---|---|---|---|

| 1 | 2 | 3 | | 5 |
|---|---|---|---|---|

| | 2 | 3 | 4 | 5 |
|---|---|---|---|---|

| 1 | | 3 | 4 | |
|---|---|---|---|---|

Nombre: _____ Fecha: _____

# FALTAN LOS NÚMEROS 1-5

Escribe los números que faltan.

| 1 | 2 |  | 4 | 5 |

| 1 | 2 | 3 |  | 5 |

|  | 2 | 3 | 4 | 5 |

| 1 |  | 3 | 4 |  |

# MISSING NUMBERS 6-10

### Write the missing numbers.

| | 7 | 8 | 9 | 10 |
|---|---|---|---|---|

| 6 | 7 | 8 | 9 | |
|---|---|---|---|---|

| 6 | 7 | | 9 | 10 |
|---|---|---|---|---|

| 6 | | 8 | | 10 |
|---|---|---|---|---|

# FALTAN LOS NÚMEROS 6-10

## Escribe los números que faltan.

| | 7 | 8 | 9 | 10 |
|---|---|---|---|---|

| 6 | 7 | 8 | 9 | |
|---|---|---|---|---|

| 6 | 7 | | 9 | 10 |
|---|---|---|---|---|

| 6 | | 8 | | 10 |
|---|---|---|---|---|

Name:_____  Date:_____

# NUMBERS

## Color, count and trace.

   one       six

 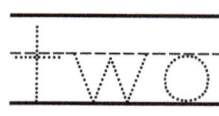  two        seven

  three      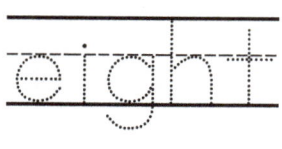  eight

 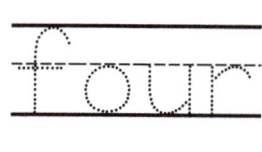  four       nine

 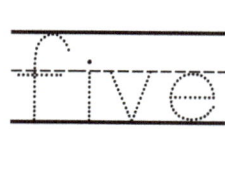  five       ten

Name: _____    Date:_____

# LOS NÚMEROS

## Colorear, contar y trazar

uno    seis

---

 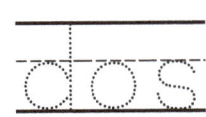     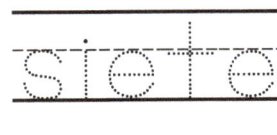

dos    siete

---

 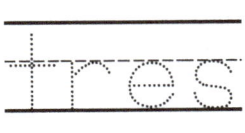     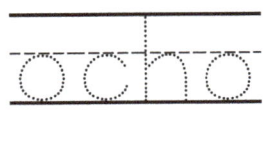

tres    ocho

---

 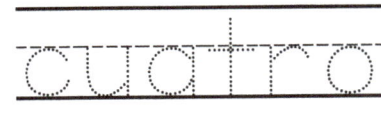    

cuatro    nueve

---

 cinco     diez

# THE TOTAL AMOUNT

**Directions: Count the objects and circle the correct number.**

**Instrucciones: Cuenta los objetos y encierra en círculo el número correcto.**

|  |  |
|---|---|
| (heart sunglasses ×4) | 5, 4, 6 |
| (suns ×6) | 2, 4, 6 |
| (mugs ×9) | 10, 9, 2 |
| (lightbulbs ×7) | 7, 4, 8 |
| (watch ×1) | 1, 5, 3 |
| (phones ×3) | 9, 3, 10 |

# ADD/SUMAR

Name: _____     Date: _____

 # Prairie Adding

**Directions: Add the butterflies and flowers.**

**Instrucciones: Sumar las mariposas y las flores.**

Name: _____          Date: _____

# NATURE SUM

Directions: Draw line to connect to the answer.

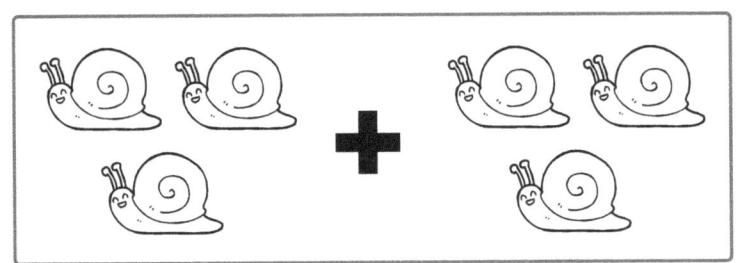

•                    •  **8**

•                    •  **6**

Instrucciones: Dibuja una línea para conectar a la respuesta.

•                    •  **3**

•                    •  **7**

Name: _____ Date: _____

# How many are there?

Count the bugs and write the number in the box.

# ¿Cuántos hay?

Cuenta el número de bichos y escribe el número en la caja.

Name: _____ Date: _____

# Adding Leaves

Count the leaves and write the number in the box.

+ =

+ =

+ =

Nombre:_____ Fecha:_____

# Sumar las hojas

Cuenta el número de hojas y escribe el número en la caja.

# Insect Counting

Add and write the answer in the box.

3 + 2 = _____

1 + 1 = _____

2 + 2 = _____

1 + 2 = _____

1 + 0 = _____

3 + 1 = _____

Nombre: _____ Fecha: _____

# Contando los insectos

Sumar y escribe el número en la caja.

$3 + 2 =$ ____

$1 + 1 =$ ____

$2 + 2 =$ ____

$1 + 2 =$ ____

$1 + 0 =$ ____

$3 + 1 =$ ____

Name: _____     Date: _____

 # WHAT'S THE SUM?

Directions: Find the sum of each equation.

Instrucciones: Suma cada ecuación.

| 4 | 1 | 4 | 5 | 2 |
|---|---|---|---|---|
| +4 | +5 | +5 | +2 | +3 |
| ___ | ___ | ___ | ___ | ___ |

| 7 | 8 | 6 | 1 | 4 |
|---|---|---|---|---|
| +3 | +2 | +2 | +3 | +2 |
| ___ | ___ | ___ | ___ | ___ |

| 4 | 8 | 6 | 2 | 5 |
|---|---|---|---|---|
| +3 | +1 | +1 | +5 | +4 |
| ___ | ___ | ___ | ___ | ___ |

# SUBTRACT/RESTAR

# Dinosaurs

**Directions**: Write an equation to match the pictures in each box. Solve the problem.

**Instrucciones**: Escribe una ecuación que corresponda a las imágenes de cada casilla. Resuelve el problema.

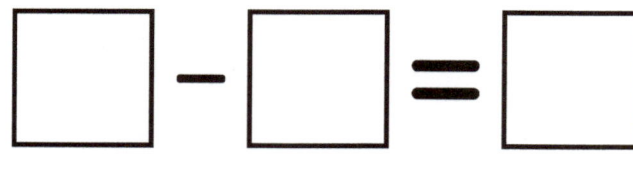

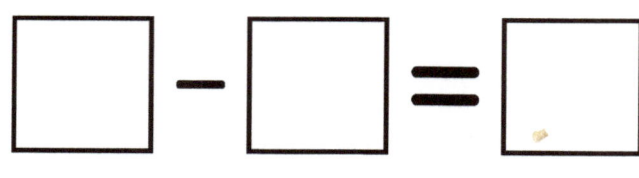

$$\square - \square = \square$$

Name: _____     Date: _____

# Koala Counting

**Directions**: Find the answer to each problem.

**Instrucciones**: Encuentra la respuesta a cada problema.

4 - 2 = _____

3 - 3 = _____

6 - 5 = _____

Name: _____     Date: _____

# MATH MINUS

Directions: Subtract each problem. Write and color the answer.
Instrucciones: Resta cada problema. Escribe y colorea la respuesta.

$$7 - 4 = \underline{\quad}$$

$$6 - 2 = \underline{\quad}$$

Name: _____    Date: _____

# SUBTRACTION 2

Subtract and write the correct answer in the box.

| | |
|---|---|
|  | 2-1 = |
|  | 3-1 = |
|  | 4-3 = |
|  | 5-1 = |
|  | 4-2 = |

Name: _____ Date: _____

# THE DIFFERENCE

Directions: Subtract the objects and draw a line to the answer.

•                    •  **2**

•                    •  **3**

Instrucciones: Resta los objetos y traza una línea a la respuesta.

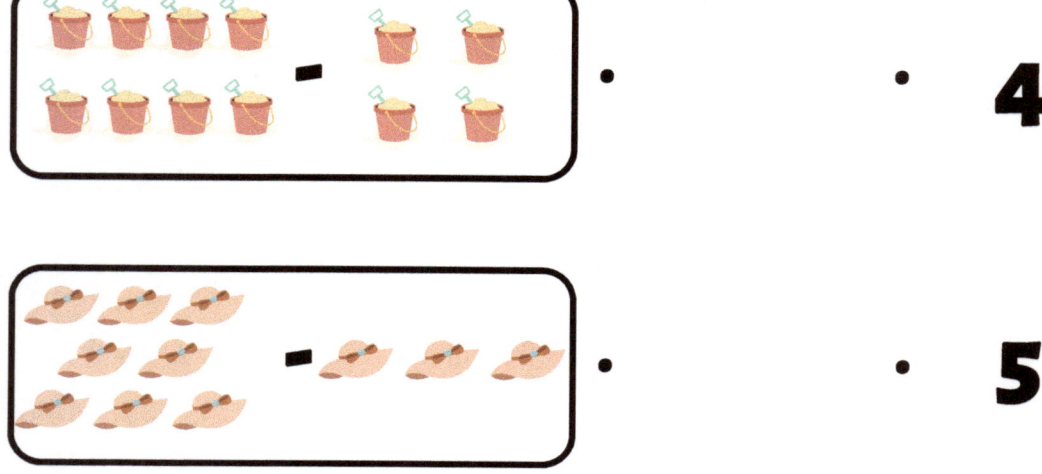

•                    •  **4**

•                    •  **5**

Name: _____  Date: _____

# SUBTRACTION

Directions: Subtract each problem.
Instrucciones: Resta cada problema.

$$\begin{array}{r} 8 \\ -\ 1 \\ \hline \end{array}\qquad \begin{array}{r} 6 \\ -\ 4 \\ \hline \end{array}\qquad \begin{array}{r} 10 \\ -\ 3 \\ \hline \end{array}\qquad \begin{array}{r} 2 \\ -\ 1 \\ \hline \end{array}$$

$$\begin{array}{r} 7 \\ -\ 0 \\ \hline \end{array}\qquad \begin{array}{r} 9 \\ -\ 3 \\ \hline \end{array}\qquad \begin{array}{r} 5 \\ -\ 1 \\ \hline \end{array}\qquad \begin{array}{r} 8 \\ -\ 5 \\ \hline \end{array}$$

# ANSWER/RESPUESTAS

Name:_____ Date:_____

## THE TOTAL AMOUNT

Directions: Count the objects and circle the correct number.
Instrucciones: Cuenta los objetos y rodea con un círculo el número correcto.

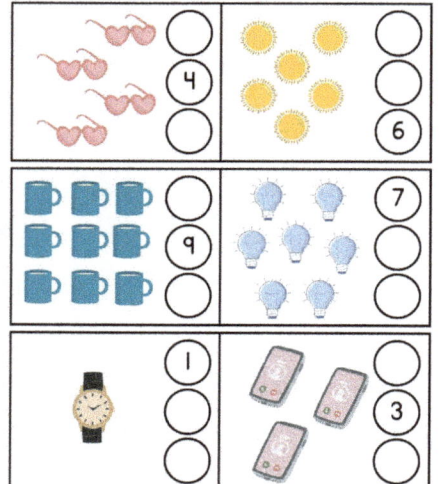

Name: _____ Date:_____

## Prairie Adding

Directions: Add the butterflies and flowers.
Instrucciones: Sumar las mariposas y las flores.

Name: _____ Date: _____

## THE SUM OF NATURE

Directions: Draw line to connect the amount of animals to the correct number.

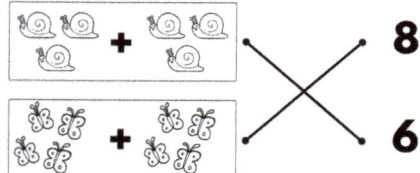

Instrucciones: Dibuja una línea para conectar la cantidad de animales con el número correcto.

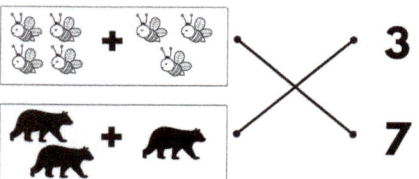

Name: _____ Date: _____

## How many are there?

Count the number of bugs in each row and write the number in the box.

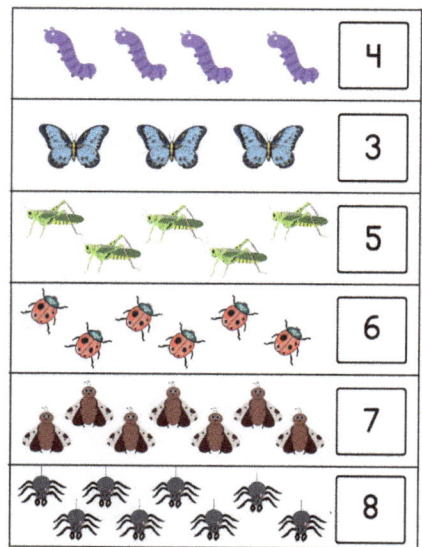

# ANSWER/RESPUESTAS

## ¿Cuántos hay?

Nombre: _____ Fecha: _____

Cuenta el número de bichos de cada fila y escribe el número en la caja.

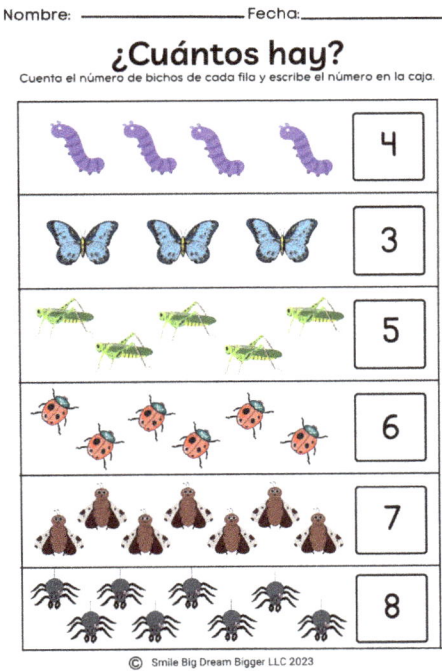

| | |
|---|---|
| (caterpillars) | 4 |
| (butterflies) | 3 |
| (grasshoppers) | 5 |
| (ladybugs) | 6 |
| (moths) | 7 |
| (spiders) | 8 |

© Smile Big Dream Bigger LLC 2023

## Adding Leaves

Name: _____ Date: _____

Count the number of leaves in each row and write the number in the box.

+ = 4

+ = 6

+ = 8

+ = 9

© Smile Big Dream Bigger LLC 2023

## Añadir hojas

Nombre: _____ Fecha: _____

Cuenta el número de hojas de cada fila y escribe el número en la caja.

+ = 4

+ = 6

+ = 8

+ = 9

© Smile Big Dream Bigger LLC 2023

## Insect Counting

Name: _____ Date: _____

Count the number of insects in each row and write the number in the box.

$3 + 2 = \underline{5}$   $1 + 1 = \underline{2}$

$2 + 2 = \underline{4}$   $1 + 2 = \underline{3}$

$1 + 0 = \underline{1}$   $3 + 1 = \underline{4}$

© Smile Big Dream Bigger LLC 2023

# ANSWER/RESPUESTAS

Nombre: _____ Fecha: _____

## Contando los insectos

Cuenta el número de insectos de cada fila y escribe el número en la caja.

3 + 2 = __5__

1 + 1 = __2__

2 + 2 = __4__

1 + 2 = __3__

1 + 0 = __1__

3 + 1 = __4__

© Smile Big Dream Bigger LLC 2023

Name: _____ Date: _____

## WHAT'S THE SUM?

Directions: Find the sum of each equation.

Instrucciones: Suma cada ecuación.

| 4 | 1 | 4 | 5 | 2 |
|---|---|---|---|---|
| +4 | +5 | +5 | +2 | +3 |

| 7 | 8 | 6 | 1 | 4 |
|---|---|---|---|---|
| +3 | +2 | +2 | +3 | +2 |

| 4 | 8 | 6 | 2 | 5 |
|---|---|---|---|---|
| +3 | +1 | +1 | +5 | +4 |

© Smile Big Dream Bigger LLC 2023

Name: _____ Date: _____

## Minus the Dinosaurs

**Directions:** Write subtraction equations to match the pictures in each box and solve.

**Instrucciones:** Escribe ecuaciones de resta que correspondan a las imágenes de cada casilla y resuélvelas.

4 − 1 = 3

6 − 2 = 4

3 − 1 = 2

© Smile Big Dream Bigger LLC 2023

Name: _____ Date: _____

## Koala Kounting

**Directions:** Find the answer to each problem.

**Instrucciones:** Encuentra la respuesta a cada problema.

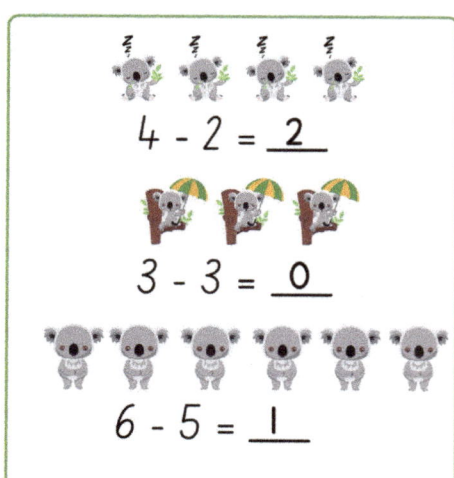

4 − 2 = __2__

3 − 3 = __0__

6 − 5 = __1__

© Smile Big Dream Bigger LLC 2023

# ANSWER/RESPUESTAS

Name: _____ Date: _____

## MATH MINUS

Directions: Subtract each problem. Write and color the answer.
Instrucciones: Resta cada problema. Escribe y colorea la respuesta.

$7 - 4 = \underline{3}$

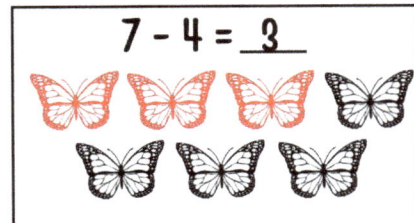

$6 - 2 = \underline{4}$

© Smile Big Dream Bigger LLC 2023

Name: _____ Date: _____

## SUBTRACTION 2

Subtract and write the correct answer in the box.

 | 2-1 = 1

 | 3-1 = 2

| 4-3 = 1

 | 5-1 = 4

 | 4-2 = 2

© Smile Big Dream Bigger LLC 2023

Name: _____ Date: _____

## THE DIFFERENCE

Directions: Subtract the objects and draw a line to the answer.

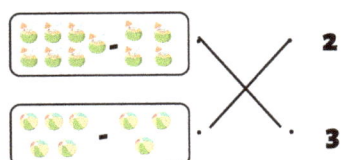

2

3

Instrucciones: Resta los objetos y traza una linea a la respuesta.

4

5

© Smile Big Dream Bigger LLC 2023

Name: _____ Date: _____

## SUBTRACTION

Directions: Subtract each problem.
Instrucciones: Resta cada problema.

$$\begin{array}{r} 8 \\ -\,1 \\ \hline \boxed{7} \end{array} \qquad \begin{array}{r} 6 \\ -\,4 \\ \hline \boxed{2} \end{array} \qquad \begin{array}{r} 10 \\ -\,3 \\ \hline \boxed{7} \end{array} \qquad \begin{array}{r} 2 \\ -\,1 \\ \hline \boxed{1} \end{array}$$

$$\begin{array}{r} 7 \\ -\,0 \\ \hline \boxed{7} \end{array} \qquad \begin{array}{r} 9 \\ -\,3 \\ \hline \boxed{6} \end{array} \qquad \begin{array}{r} 5 \\ -\,1 \\ \hline \boxed{4} \end{array} \qquad \begin{array}{r} 8 \\ -\,5 \\ \hline \boxed{3} \end{array}$$

© Smile Big Dream Bigger LLC 2023